PINGREE FARMS

CHANGING SPACES

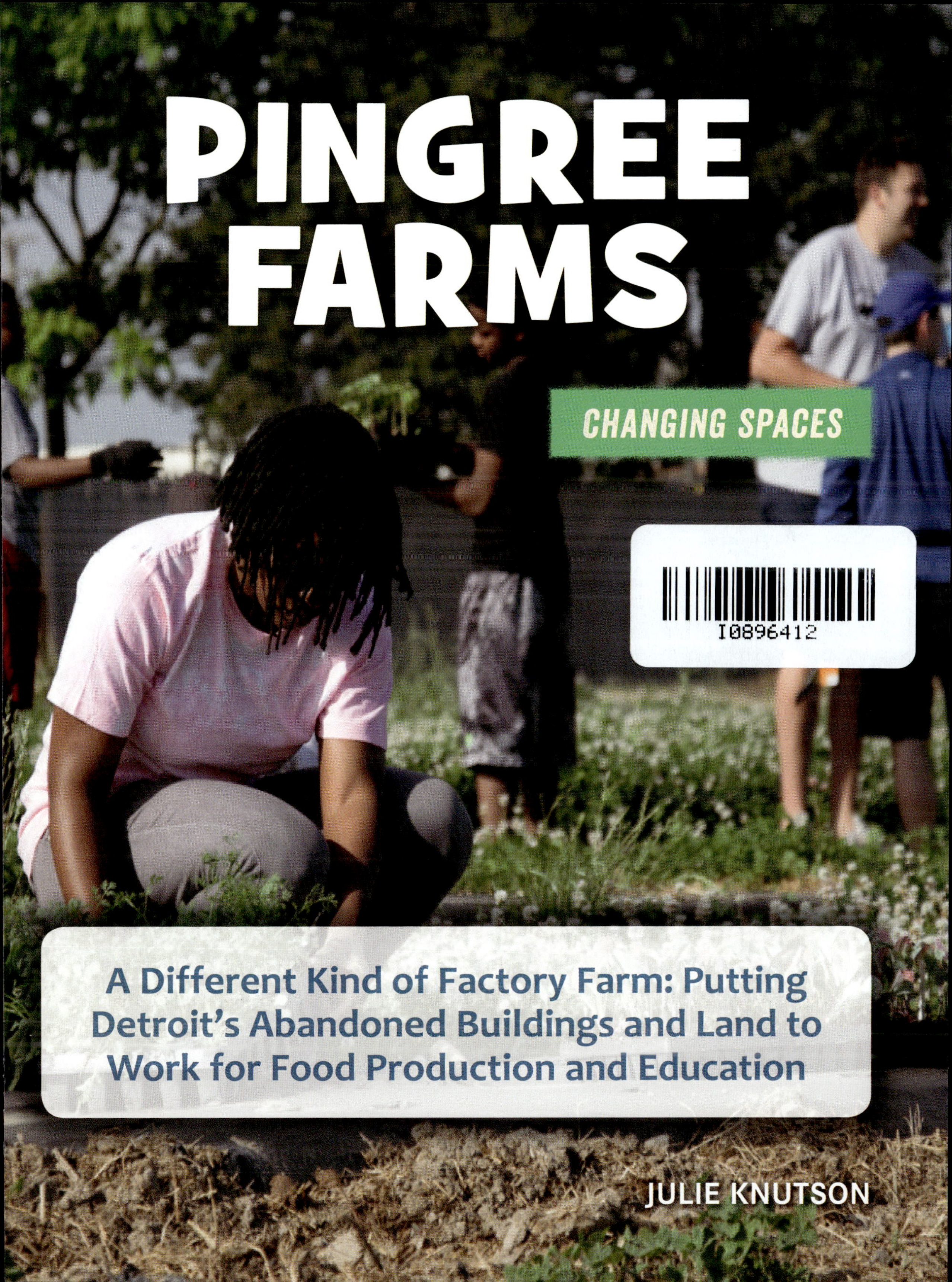

A Different Kind of Factory Farm: Putting Detroit's Abandoned Buildings and Land to Work for Food Production and Education

JULIE KNUTSON

Cherry Lake Press

Published in the United States of America by Cherry Lake Publishing Group
Ann Arbor, Michigan
www.cherrylakepublishing.com

Reading Adviser: Marla Conn, MS, Ed., Literacy specialist, Read-Ability, Inc.
Photo Credits: © Courtesy of Pingree Farms, cover, 1, 5, 6, 20, 22, 23, 25, 26, 28; © Everett Historical/Shutterstock.com, 9; © Andrey Bayda/Shutterstock.com, 10; © stockddvideo/Shutterstock.com, 14; © Mathias Berlin/Shutterstock.com, 17; © alisafarov/Shutterstock.com, 19

Cherry Lake Press is an imprint of Cherry Lake Publishing Group.

Library of Congress Cataloging-in-Publication Data
Names: Knutson, Julie, author.
Title: Pingree Farms / by Julie Knutson.
Description: Ann Arbor, Michigan : Cherry Lake Publishing, [2021]. | Series: Changing spaces | Includes index. | Audience: Grades 4-6 | Summary: "Learn more about how Detroit's Pingree Farms went from vacant industrial city block to vibrant urban farm and community center. Explore the logistics of repurposing the land and meet the people who made it happen. The book showcases a range of 21st century skills—from "Flexibility & Adaptation" to "Creativity & Innovation"—and shows how moving away from a tear-down culture towards one of reuse helps tackle a host of critical challenges facing our planet and population. Thought-provoking questions and hands-on activities encourage the development of critical life skills and social emotional growth. Books in this series include table of contents, glossary of key words, index, author biography, sidebars, and infographics"—Provided by publisher.
Identifiers: LCCN 2020006386 (print) | LCCN 2020006387 (ebook) | ISBN 9781534169050 (hardcover) | ISBN 9781534170735 (paperback) | ISBN 9781534172579 (pdf) | ISBN 9781534174412 (ebook)
Subjects: LCSH: Pingree Farms (Detroit, Mich.)—Juvenile literature. | Urban agriculture—Michigan—Detroit—Juvenile literature. | Deindustrialization—Social aspects—Michigan—Detroit—Juvenile literature.
Classification: LCC S451.M5 K59 2021 (print) | LCC S451.M5 (ebook) | DDC 630.9774/34—dc23
LC record available at https://lccn.loc.gov/2020006386
LC ebook record available at https://lccn.loc.gov/2020006387

Cherry Lake Publishing Group would like to acknowledge the work of the Partnership for 21st Century Learning, a Network of Battelle for Kids. Please visit http://www.battelleforkids.org/networks/p21 for more information.

Printed in the United States of America
Corporate Graphics

With thanks to Juana Woodson and Jim Green

ABOUT THE AUTHOR

Julie Knutson is an author/educator fascinated by the endless ways in which people and communities can transform old spaces for new uses. She lives in Illinois with her husband, son, and extremely energetic border collie.

TABLE OF CONTENTS

Adaptive Reuse—More Than Meets the Eye

Essential question: *How can agriculture revitalize urban communities?*

On most days, the sights and sounds of Milton Manufacturing in Detroit, Michigan, are like those of any other workplace. In the office, fingers glide across keyboards, typing emails and entering orders. Phones ring, and employees answer and announce their names. On the factory floor, machines whir. Metals get melted and stamped into tidy shapes. Parts get packed into cardboard boxes, which are stacked for shipment to all corners of the globe.

The spring chicks are cared for by Pingree Farm workers and students.

But for a few days each year, the company sounds a bit less predictable . . . and a bit chirpier. Why?

Once a year, the U.S. Postal Service delivers a new group of **avian** residents. These freshly hatched spring chickens and turkeys live on Pingree Farms, an agricultural operation sponsored by the company on adjacent land—well within chirping distance. “Everybody can hear the chicks chirping and cheeping,” laughs Juana Woodson, Milton Manufacturing’s property and farm manager.

This is *definitely* the day you want it to be “Take Your Child to Work Day.”

Pingree Farms has 15 acres (6 hectares) of urban farm fields.

Hazen and Frances Pingree, Champions of Urban Agriculture

Pingree Farms is named after former Detroit mayor Hazen Pingree. In the late 19th century, during a major economic **depression**, *Pingree urged citizens to plant their own "potato patches" and vegetable gardens on unused land. While Hazen is often credited with developing this innovative path to food independence, the idea actually originated with his wife, Frances.*

At Pingree Farms, chickens, goats, cows, and rabbits live in **adaptively reused** machine shops, warehouses, and garages. Reclaimed parking lots provide space for the animals to roam. Empty lots that surround the facility have been turned into farming space. In the summer, community members are free to pick squash, tomatoes, and peppers from these gardens.

Car Parks 2.0

Designers worldwide are visualizing new, greener uses for unused urban spaces. In France, the design firm Studio NAB sees the potential to transform parking lots into greenhouses and orchards. In New York, grocery store rooftops are being used as growing zones by companies like Gotham Greens. These creative agricultural enterprises stimulate local economies and reduce carbon emissions, thanks to fewer miles food has to travel.

How did unused land in this corner of the Motor City find new life as a farm? How did the idea to use it as a place to grow food for the community and provide agricultural education for local youth emerge? Let's journey to the corner of I-75 and 7 Mile Road to learn more about the past, present, and future of this unique experiment in urban agriculture.

CHAPTER 1

An Economic Powerhouse on the Great Lakes

Milton Manufacturing opened in 1946 in one of America's most economically thriving cities. With strategic access to the Erie Canal, Detroit had been a portal for trade and shipping for more than a century. And as a distribution hub for industrial materials and machinery, Detroit was the perfect home for the growing auto industry in the 1890s and early 20th century.

The auto industry employed an enormous number of workers.

For decades during the 20th century, Detroit was one of America's most vibrant cities.

GM: The Motor City's Most Profitable Carmaker

From the 1920s on, General Motors (also known as GM) promised a car for "every purse and purpose," from Chevrolets to Oldsmobiles to Cadillacs. Between 1955 and 1974, the company topped Fortune *magazine's list of the 500 most profitable U.S. companies. Facing competition, a changing market, and management issues, the company's fortunes began to decline in the 1970s and 1980s.*

Both Ford Motor Company and General Motors were headquartered in Detroit, leading car manufacturing to become the engine that drove the city's economy. Supporting industries, like auto parts makers that produced everything from fasteners to grommets, sprung up across the city. Other retail and service industries—restaurants, movie theaters, pharmacies—moved in to meet the needs of new residents drawn by the boom. By the middle of the 20th century, Detroit was the nation's fifth-largest city. It led the U.S. economy and accounted for one-sixth of the country's employment.

Immigration to Detroit

Detroit's auto industry was a major **pull factor** that attracted people to the city from other parts of the United States and the world. Drawn by the promise of good jobs, housing, and education, many African Americans moved to Detroit from the South during the period known as the Great Migration. Other northern cities—New York, Chicago, Philadelphia, Cleveland, and Baltimore—similarly drew large numbers of African Americans who were leaving an increasingly violent and segregated South governed by **Jim Crow laws**.

Though the environment in Detroit was relatively more favorable, newly arrived African Americans still encountered racism and discrimination, particularly in housing.

Ford Motor Company's $5-a-day wages and international recruiting efforts also proved a magnet to people living outside of the United States. Immigrants came to Detroit from around the globe, making it one of America's most diverse melting pots. By 1930, 30 percent of the city's population was foreign-born.

Robotic production equipment reduced the need for human auto workers.

But the economy wasn't **diversified**, and Detroit's prosperity didn't last. In the second half of the 20th century, large economic shifts led many of the city's businesses to relocate or close their doors. The growth of highways sparked more roads into the suburbs, where many companies moved their headquarters. The once-dominant American auto industry faced increasing international competition and a shrinking share of the market. **Automated** processes made human labor less necessary for many factory jobs. Other employers were lured **offshore** by the promise of less expensive operating costs. All of these forces resulted in a loss of businesses and a loss of jobs within the city limits.

CHAPTER 2

Declining Fortunes

Many overlapping economic and social factors caused the exodus of industry from Detroit. In the second half of the 20th century, the city suffered massive population exit. It wilted from nearly 2 million people in 1950 to 672,000 in 2018, a more than 60 percent loss of population. Many residents who had the resources to do so left the city's urban center and moved to the suburbs. This loss of people and businesses meant that tax **revenues** for Detroit's public schools and important services suffered. By the 2010s, there were more than 70,000 abandoned structures in the city.

Abandoned spaces in Detroit included factories, office buildings, homes, theaters, and churches.

Five Days of Rebellion in July 1967

Because of housing discrimination and segregation, racial tension simmered in Detroit for decades. In June 1943, race riots fueled by rumor and suspicion swept the city for 3 days. Thirty-four people, most of them African American, were killed.

In 1967, many African Americans in the city rebelled against oppressive conditions and racial profiling by the city's police force. The uprising started in late July, when police raided an unlicensed bar in a black community. A crowd gathered in protest, which soon grew in intensity. Buildings were burned, shops were looted, and law enforcement brought in tanks and heavily armed officers to contain the violence.

Over the course of 5 days, 43 people died and over 1,200 were injured. More than 2,000 buildings were destroyed. Much of the city was not rebuilt after the uprising.

By the 2010s, you could buy abandoned homes in Detroit for $500.

This decline of population and infrastructure brought a new round of challenges. Empty properties attracted illegal activities. Milton Manufacturing managed to survive by diversifying the range of products its shops made. But in the neighborhood surrounding it, many of the 127 homes stood vacant. The 21 other businesses that lined the street in the late 1960s were packing up and calling it quits. In their place, “chop shops” for stolen cars cropped-up in unused garages, and drug houses moved into some of the abandoned homes.

With so much empty space, Detroit lends itself well to urban farming.

When a City Declares Bankruptcy

When more money is spent than received, a person or organization goes into debt. This means they have bills from a lender or supplier that can't be paid. For years, Detroit had a supply of incoming cash from tax revenue. The city used this money to pay for everything from road construction to streetlights. But those dollars evaporated when businesses and residents left. The city's budget didn't adjust accordingly, and in 2013, Detroit declared **bankruptcy**.

By the late 1980s, criminal activity in the area surrounding Milton Manufacturing made many customers reluctant to approach it. But rather than leave, Milton's owners, Jim and Shelly Green, took a different approach. They started buying up the plots around them. They tore down decaying structures and let grass grow where houses and machine shops once stood.

Then, Jim Green—who studied agricultural engineering in college and had long aspired to be a farmer—had a seed of an idea. What if they planted corn and beans on one or two of the lots?

Pingree Farms introduces families and kids to all aspects of planting, cultivating, and farming land.

Planting Day encourages community members to take part in the farm's activities.

The idea grew. Soon, the family was planting seedlings by the thousands. Curious neighbors inquired, "How can we help?" In June 2011, Pingree Farms hosted its first Planting Day, welcoming 100-plus volunteers for a morning of work. A mariachi band strummed as holes were dug and seeds were planted. A cookout and an afternoon of festivities followed. A new space for community engagement started to take root, and anyone driving by was welcome to share in the bounty of crops.

CHAPTER 3

Growing into the Future

"Pingree is just . . . fun. I get to be whoever I want to be."

"When I leave my home to go to Pingree, I'm just going to my second home."

"[At Pingree] I've learned to never give up; to go for things out of my comfort zone."

These reflections were offered by three members of Pingree's 4-H Club, all of whom participated in the 2019 Wayne County Fair. The club draws students from 6 to 18 years old and provides a space for them to practice leadership skills while pursuing their

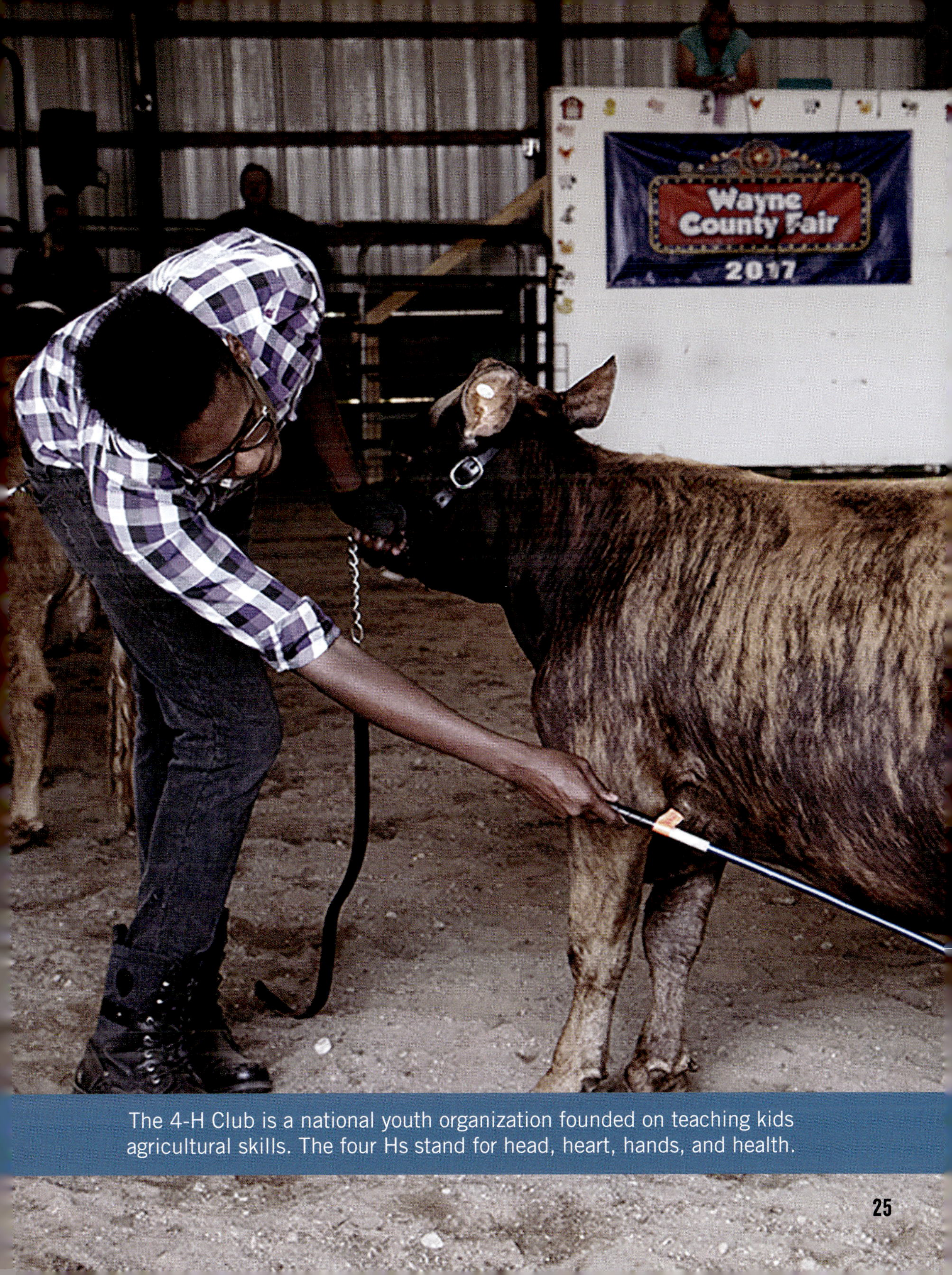

The 4-H Club is a national youth organization founded on teaching kids agricultural skills. The four Hs stand for head, heart, hands, and health.

Student programs foster interest in science and agriculture—as well as teaching leadership skills.

interest in animal science. At the fair, the kids show rabbits, cows, and goats and participate in other contests, ranging from baking to art to archery. Club members also operate a USDA-approved petting zoo for special events in the city, like open houses and planting days.

Detroit Ag City?

From Keep Growing Detroit to D-Town Farm to The Greening of Detroit, passionate city residents are collaborating to build an infrastructure that supports local food systems. According to the Urban Land Institute, "More than 23,000 Detroit residents participate in urban gardening, and the city boasts more than 1,500 individual gardens."

Pingree expanded to include student programs in 2013. Because of limited access to farm animals within the city, a local 4-H Club was disbanding. Jim and Shelly Green saw an opportunity. First, they introduced cows to the Pingree property, working with a local community college to make it a true learning center. As they expanded their flocks and herds, unused buildings on the industrial side of the street provided the perfect shelters. Keeping livestock on this nonresidential side of the street was essential for legal reasons,

Goat yoga is one of Pingree Farms' most popular fundraising events.

as it allowed the site to obtain laboratory status. Vacant old office buildings, machine shops, and garages were adapted for housing not just animals, but also a one-room schoolhouse.

Today, Pingree supports learning for students of all ages. Master Gardeners with Michigan State University experiment with **viticulture**, growing grapes on the property. These same green-thumbed adults coach 4-H students in **horticultural** science. This year, a new partnership will allow 15 students from a local school the chance to spend 1 day a week on the farm. The Greens and farm manager Juana Woodson hope to build additional programs with area educators in the coming years. They also hope to add orchards to the site, which would allow area residents to stop in to pick apples, pears, and plums.

With nearly 10 years of grassroots efforts behind them, the folks at Pingree Farms keep looking toward the future, seeing room for growth and new partnerships on the shared road to a greener Detroit.

Ag Education for a New Generation

In developing programs for youth, Pingree Farms has looked within and beyond Detroit for ideas. For Jim Green, one unique site of inspiration is the Chicago High School for Agricultural Sciences (CHSAS). For 35 years, CHSAS has prepared graduates for careers in animal science, horticulture, biotech, and agricultural economics. Supported by a 39-acre (16 ha) farm and countless community partnerships, the school equips its future farmers with the immediate and long-term skills needed to create **sustainable** *food systems. Jim hopes to see school districts in Detroit and elsewhere launch similar programs.*

Extend Your Learning

Small-scale backyard and community gardens provide a way for people of all ages to fight climate change. They reduce the number of miles that food travels and decrease carbon emissions. Plants also capture carbon dioxide in the soil, removing it from the atmosphere.

As food activist Acadia Tucker notes, there are 40 million acres (16 million ha) of unused green space in the United States that could be farmed. Research what crops grow best in your climate region and soil conditions. Search for land at home and in your community for potential gardens. Develop a plan that addresses *who, what, when, where, why,* and *how* of a potential garden.

Present this plan to adult allies—parents, grandparents, teachers, school and community leaders—and ask for help and resources from them to grow food on these unused plots.

Glossary

adaptively reused (uh-DAP-tiv-lee ree-YOOZD) to reuse a structure for a purpose other than that which it was originally designed to serve

automated (AW-tuh-mayt-id) the use of machines rather than people to do jobs, especially in factories

avian (AY-vee-uhn) of or relating to birds

bankruptcy (BANGK-ruhpt-see) a legal process required for people or companies unable to pay their debts

depression (dih-PRESH-uhn) a period of economic decline

diversified (dih-VURS-ih-fied) having many different industries, not just one

horticultural (hor-tih-KUHL-chur-uhl) the cultivation of plants and gardens

Jim Crow laws (JIM KROH LAWZ) laws in the United States that legalized segregation from the late-19th to mid-20th centuries and set different rules for black and white Americans

offshore (AWF-shor) outside the country

pull factor (PUL FAK-tur) something that draws or "pulls" a person to migrate to a certain place

revenues (REV-uh-nooz) incoming sources of money

sustainable (suh-STAY-nuh-buhl) done in a way that can be continued and that doesn't use up natural resources

viticulture (VIH-tih-kuhl-chur) the cultivation of grapes, especially for wine making

INDEX